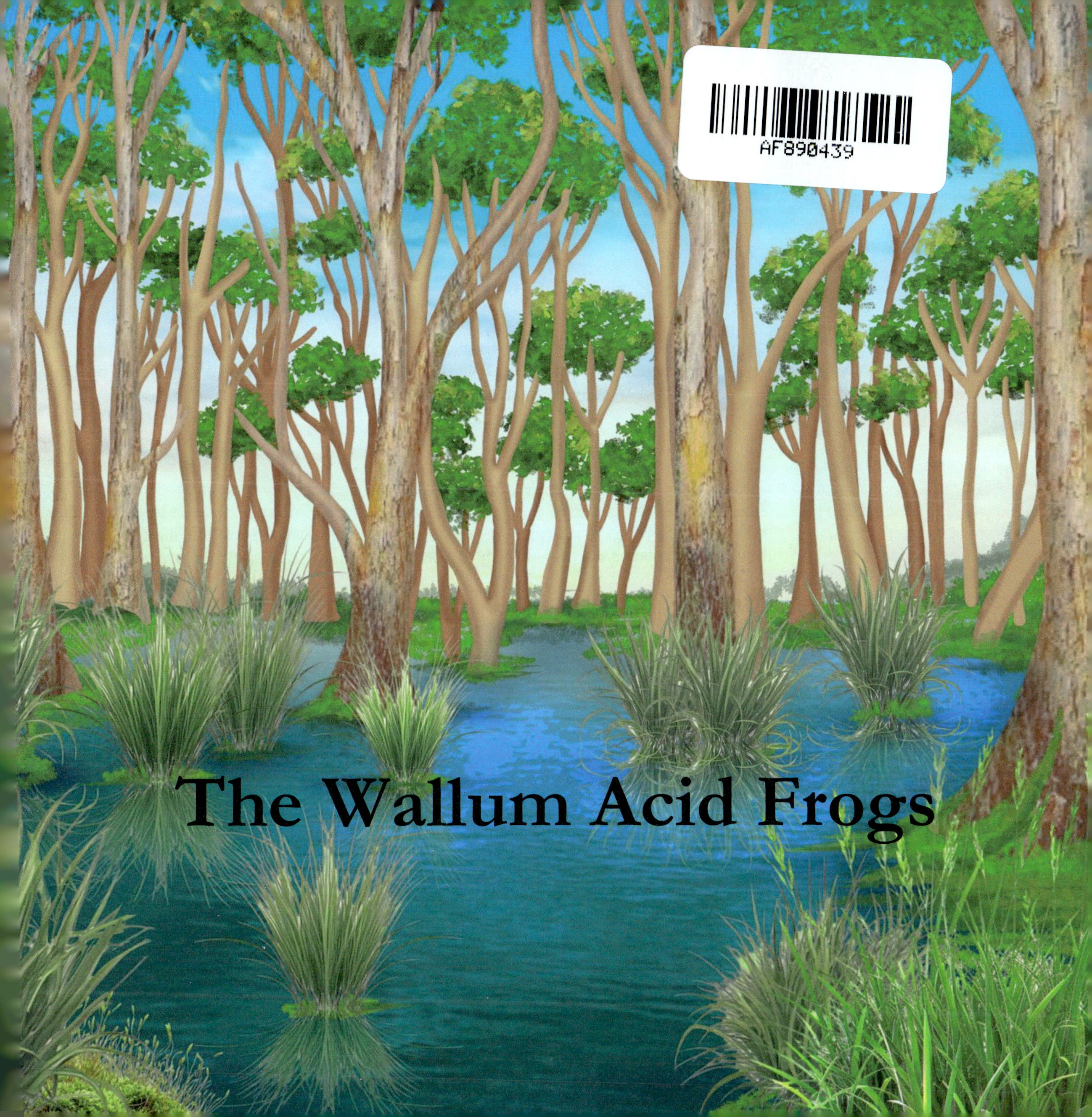
The Wallum Acid Frogs

Illustrations by Sidra Aftab @epicillustrator - Fiverr

ISBN: 978-1-7641194-6-7 (Paperback)

First edition June 2025
A CONSERVATION CLUB PUBLICATION
www.conservationclub.com.au

NATIONAL LIBRARY OF AUSTRALIA

A catalogue record for this book is available from the National Library of Australia

In the spirit of reconciliation, Conservation Club acknowledges the Traditional Custodians of Country throughout Australia and their connection to land, sea, and community. We pay our respects to their Elders, past and present, and extend that respect to all Aboriginal and Torres Strait Islander peoples.

To my grandchildren, Samuel, Jameson, Harry and Audrey.

I hope that you all have the same sense of wonder and love of the environment that I enjoy as you grow into bigger children, young adults and into adulthood, and also that you have the opportunity to explore many different environments, including the Wallum during your lifetimes.

Welcome to the Wallum!

Deep in the coastal lands of eastern Australia lies a wild, watery world called **The Wallum**. This special place is filled with swamps, marshes, heathlands, and dark, acidic pools. It may not look like much at first, but it's full of secrets—and at least four of them hop on tiny legs!

<u>Are you ready to leap in</u>?

A handful of these secrets are the acid frogs - rare, sometimes colourful, and endangered frogs, which are found nowhere else in the world!

In this book, you'll meet four of these special "acid frogs" who call The Wallum home.

- The Wallum sedgefrog
- The Wallum froglet
- The Cooloola sedgefrog
- The Wallum rocketfrog

Can you guess which one is the Wallum rocketfrog?

Whoo--oo--sh.....

What is The Wallum?

The Wallum is a special kind of habitat found along the sandy coastal plains of Queensland and Northern New South Wales. It is made up of swamps and wetlands, often with acidic waters.

Why is the water so acidic?

It's because of the tannins from nearby plants, like 'tea trees', that stain the water brown—almost like tea!

Many animals can't survive in this sour soup, but these special frogs have adapted to live here.......and now, they need our help.

Wallum Sedge Frog (Litoria olongburensis)

Meet the **Wallum sedge frog**

This tiny frog is only about 3 cm long - about the size of a standard paperclip – actually the standard paperclip is 0.5cm longer!

....Get out a paperclip and measure how big that is! I think you will be surprised...

Wallum sedge frogs have bright green or bronze coloured backs, with a dark brown stripe from the tip of the snout to the side. There is a white stripe under the eye to the arm, which becomes spots as it reaches the side.

They're active both day and night, and love hiding among reeds and grasses near still water.

They have a loud, high-pitched call that sounds like a cricket! It's often the only way scientists can find them.

This Wallum acid frog can live in very acidic waters—something most frogs can't do. It lays its eggs in shallow puddles and ponds, where the tadpoles hatch and grow.

In this picture you can see five stages that a frog goes through when it is growing, from the eggs to tadpoles, to growing both back and front legs and, eventually, losing its tail.

Have you ever collected tadpoles and watched them develop? Perhaps this is something you could try!

These Wallum “acid frogs” are sensitive to pollution and need clean, acidic water to breed.

Without healthy wetlands, their numbers will continue to shrink.

Sadly, their habitat is shrinking due to housing developments and pollution.

Without clean, acidic wetlands, the Wallum acid frogs are losing their home.

What can you see in this picture that might be affecting the Wallum wetlands?

Bin it, don't fling it!!

This is a reminder that it is so important that you do not leave any rubbish in any environments that you visit!

Wallum Rocket Frog (Litoria freycineti)

Get ready to meet a frog with serious spring power!

The **Wallum rocket frog** is built for speed and distance and can jump around two metres in a single leap!

It is called a rocket frog for a reason but perhaps not the reason you are thinking of!

Wallum rocket frogs have strong legs and long toes that help them leap great distances. When startled, they take off like a rocket—but they are called a rocket frog because of their overall shape and pointy snout!

These frogs are pale brown with dark markings and can be hard to spot unless they move.

Unlike many other frogs, rocket frogs don’t lay their eggs in water. Instead, they tuck their eggs into moist soil or moss near the water’s edge.

When the tadpoles hatch, they wriggle their way to water.

Habitat destruction is the biggest threat to this speedy frog. When wetlands are drained or filled in, the rocket frog has nowhere to lay its eggs.

Wallum Froglet (Crinia tinnula)

The tiniest of the bunch is the **Wallum froglet** - it's only about 2 cm long—smaller than a grape!

Despite its size, it makes a big noise.

The Wallum froglet's call sounds like a tiny bell ringing: *"tink tink tink"!* That's why its scientific name is *tinnula*, which means "tiny bell."

The Wallum froglet lives in Wallum swamps and shallow pools, where it lays eggs in small clusters.

Unlike most frogs, the Wallum froglet doesn’t hop much—it prefers to crawl!

Because they are so small and quiet, these froglets are often overlooked. But their wetlands are disappearing, and they are in danger too.

Cooloola Sedge Frog (Litoria cooloolensis)

This small frog grows up to 3cm in body length. Its back can be bright green or brown with small brown spots. Some have a dark stripe from nose to eye, and a white stripe from under the eye to the arm.

The belly is cream, the eyes are gold with sideways pupils, and the backs of the legs are orange with a purple-brown stripe.

Its toes are webbed for swimming, with big round pads for climbing.

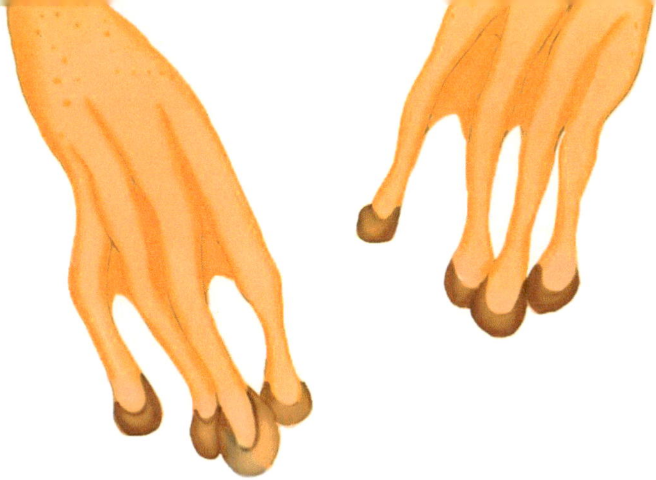

This frog lays eggs in clusters under the water, attached to reeds in ponds, swamps, and lakes.

Tadpoles grow up to 5.5 cm long, gold with dark spots, and swim in shallow water.

No one knows how long they take to develop into frogs. They breed from spring to autumn after rain.

Why Are These Frogs Endangered?

All four of these frogs are **endangered** or **vulnerable**, which means they're at risk of disappearing forever.

Here's why:

- **Habitat loss**: Wetlands are drained for farms and houses.
- **Pollution**: Chemicals in the water can kill frogs and tadpoles.
- **Climate change**: Rising temperatures and changing rain patterns affect breeding.
- **Invasive species**: Animals like cane toads, fish, and feral pigs eat frogs or compete with them.
- **Disease**: A deadly fungus called chytrid affects frog skin and can kill them.

Frogs are like nature's alarm bells. When frogs are in trouble, it means that the environment is in trouble too.

Frogs might be small, but they play BIG roles in the environment:

- They eat mosquitoes, flies, and other insects.
- They are food for snakes, birds, and fish.
- Their skin can absorb water and even breathe—scientists study it for clues to new medicines.
- Their eggs and tadpoles help keep wetlands clean and balanced.

If frogs disappear, the whole ecosystem suffers. Saving frogs means saving the places they (and we!) live in.

Even kids can be frog heroes!

Here's how:

- **Learn more**: Read about frogs and share what you know.
- **Keep wetlands clean**: Don't dump waste into ponds or creeks.
- **Plant native plants**: Frogs need healthy, shady wetlands.
- **Be careful with chemicals**: At home and in gardens.
- **Stay on paths**: In parks and nature reserves to avoid damaging habitats.
- **Listen and report**: If you hear frog calls in the wild, tell a local frogwatch group!

You can even become a **citizen scientist** and help count frogs in your area!

The future of the Wallum frogs

Scientists, rangers, and communities are working hard to protect these special frogs. They’re restoring wetlands, studying frog diseases, and creating protected areas.

But they can’t do it alone.

Frogs may be quiet, hidden, and small—but they need a loud voice. Your voice.

Will you help speak up for the frogs of the Wallum?

A Thank You from the Frogs

Thank you for taking the time to learn about a few of Australia’s incredible frogs. They may be small, but they’re full of life, colour, and surprises.

If we all do our part, we can make sure these froggy friends have a safe and splashy future.

Until next time, keep hopping forward!

ABOUT THE AUTHOR

Meet Elysia Rainwood, an artist and writer with a passion for conservation and environmental protection, particularly the preservation of the diverse animals inhabiting our wonderful Australian continent.

Growing up Elysia always had an eye for detail, being very aware of such things as seasonal changes in the various environments she saw, from the depths of the lush, green rainforests to the sun-kissed beaches of the outer Great Barrier Reef islands, her adventures fueled a desire to share the stories of the vulnerable and endangered.

Elysia writes non-fiction books and stories that resonate with the spirit of the tiniest of creatures and soar with the majesty of the wedge-tailed eagle. She likes to focus on the less well-known species to ensure that they and their habitats are not overlooked.

Elysia Rainwood's mission is clear:

To enchant young minds with the wonders of Australia's wildlife, fostering a generation of conservationists. Through interesting facts, engaging and reasonably accurate illustrations, she hopes to ignite a spark in the hearts of 4 to 10-year-old readers, inspiring them to protect the precious creatures that share our planet.

Along the way, Elysia hopes to also engage parents, care-givers, and educators who can pass on this information.